Weird World of Animals

Animals That Lose Their Limbs

By Mary Griffin

Gareth Stevens Publishing

Please visit our website, www.garethstevens.com. For a free color catalog of all our high-quality books, call toll free 1-800-542-2595 or fax 1-877-542-2596.

Names: Griffin, Mary, 1978-.
Title: Animals that lose their limbs / Mary Griffin.
Description: Buffalo, NY : Gareth Stevens Publishing, 2026. | Series: Weird world of animals | Includes glossary and index.
Identifiers: ISBN 9781482471717 (pbk.) | ISBN 9781482471724 (library bound) | ISBN 9781482471731 (ebook)
Subjects: LCSH: Animal defenses–Juvenile literature. | Animal defenses–Adaptation–Juvenile literature. | Animal behavior–Juvenile literature.
Classification: LCC QL759.G75 2026 | DDC 591.47–dc23

Published in 2026 by
Gareth Stevens Publishing
2544 Clinton Street
Buffalo, NY 14224

Cover Designer: Nicholas Switalski
Interior Designer: Rachel Rising
Editor: Therese Shea

Photo credits: Cover, p. 1 Stubblefield Photography/Shutterstock.com; Cover, pp. 1, 3-24 Tunatura/Shutterstock.com; p. 5 Awei/Shutterstock.com; p. 5 (inset) Md Aziman/Shutterstock.com; p. 7 Lauren Suryanata/Shutterstock.com; p. 8 Eric Isselee/Shutterstock.com; p. 11 Evgeny Shaposhnikov/Shutterstock.com; p. 12 renuflora/Shutterstock.com; p. 13 axolotlowner/Shutterstock.com; p. 15 Allgord/Shutterstock.com; p. 15 (inset) Shahidphotographer/Shutterstock.com; p. 16 photobeps/Shutterstock.com; p. 17 JMetz/Shutterstock.com; p.19 simone tognon/Shutterstock.com; p. 21 Vojce/Shutterstock.com.

Printed in the United States of America

CPSIA compliance information: Batch #CSGS26: For further information contact Gareth Stevens, at 1-800-542-2595.

CONTENTS

*Words in the glossary appear in **bold** type the first time they are used in the text*

Amazing Adaptations

An adaptation is the way nature makes animals and plants better able to live in their **habitats**. An adaptation can be a body part or a way of acting. For example, brown birds can blend in with their forest habitats. Plants may grow toward the sun to take in more sunlight for **energy** to grow.

Some adaptations are a bit stranger. You might even think they're weird. For example, some animals can lose their limbs, or arms and legs. That's a way they survive! Read on to learn more.

Some spiders drop a leg if they're stung there by a bee or wasp. This spider is missing two legs.

That's WEIRD!

Some bugs such as crickets drop a leg to **distract** predators. The predator looks at the leg while the bug gets away!

Newts' Neat Trick

About 40 species, or kinds, of salamanders are called newts. All newts have a long, thin body and a flat tail. Most don't even reach 8 inches (20 cm) long. They live in the Northern **Hemisphere**. Some species spend much of their lives in water. Others live on land. Most lay eggs in the water, though.

Newts have many predators, including birds, foxes, hedgehogs, snakes, and cats. Newts have several ways to keep from being eaten. One of the best is being able to lose and regrow limbs!

This species of newt is called a crocodile newt. Its coloring acts as **camouflage**. That's an adaptation too.

CROCODILE NEWT

That's WEIRD!

Some newts have super-poisonous skin. A rough-skinned newt's skin is so poisonous it kills most animals that eat it!

If a predator bites off a newt's leg or tail, it can grow a new one. Scientists have studied this adaptation and the **gene** that controls it. Special cells gather where the limb was bitten off. The cells change to have new purposes. Over time, they build the bones, muscles, skin, and more needed for a limb.

A newt can regrow the same body part again and again. And it's not just limbs. Newts can grow new ears, eyes, hearts, and brains too!

It takes a few months for newts to regrow a limb.

Scientists are amazed the newt's body can rewrite cells' instructions.

How a Newt Regrows a Limb

1. New skin forms over the wounded area.
2. A group of cells called the blastema forms at the wound.
3. The cells' original purpose in other parts of the body is erased.
4. The cells are given new directions.
5. The cells form a new limb.

The Axolotl's Abilities

Axolotls are another species of salamander. They have a long tail, short limbs, and a fin down their back. They have feathery **gills** on each side of their head. They often have a dark coloring, but some are white or pink.

Axolotls live in water their entire lives. Their predators include storks, herons, and large fish. If one of these predators gets hold of an axolotl's limb, this salamander can survive without it—and even regrow it.

Axolotls stay in their larva form their whole lives. They don't change as other salamanders do.

That's WEIRD!

Axolotls have no teeth. They use a powerful sucking force to eat worms, bugs, and small fish.

The axolotl can regenerate, or regrow, its limbs, tail, gills, lungs, brain, jaws, **spinal cord**, and more! Scientists have been studying the axolotl for many years. They raise them in labs, so they can study their adaptations.

Sadly, even the superpower of regeneration can't save the axolotl population in the wild. They're found only in a couple of small bodies of water in Mexico City. They're in danger of dying out. Some think there are fewer than 100 left in the wild.

This axolotl is regrowing its front leg and tail.

SEA STARS' SUPERPOWERS

Sea stars are **marine** animals without a backbone. There are around 1,600 sea star species. They live in every ocean. The largest can grow to be 2 feet (61 cm) wide! Most are much smaller. They can be many colors.

Sea stars have arms called rays that grow from the center of their body. On the underside of the rays are tiny tube feet that act a bit like legs. They help the sea star move. Other parts on sea star rays act like **suction cups** to grab surfaces.

Some people call sea stars "starfish," but they aren't fish.

That's WEIRD!

Most sea stars have five rays, or arms. Some have 50!

Many animals eat sea stars, including crabs, lobsters, sea gulls, turtles, otters, and other sea stars. If one of these creatures eats a sea star's ray, the sea star can grow a new one. But the sea star has another handy adaptation. It can detach, or let go of, a ray to get away!

For some species, the detached ray can grow into another sea star. That sea star will be **identical** to the sea star the ray came from.

This photo shows a sea star growing a new ray!

That's WEIRD!

When sea stars eat their prey, their whole stomach comes out of their mouth!

Sea Spiders' Legs and More

A sea spider isn't a spider. It's an **arthropod** that often lives on the ocean floor. It has four pairs of long legs.

Scientists long knew sea spiders could regrow parts of their legs. They now know young sea spiders can regrow the whole back half of their body. Their regeneration isn't always perfect. Some regrow fewer legs. Adult sea spiders can't regenerate the back half of their body. Scientists think this is because young sea spiders **molt** as they grow, while adults have stopped molting and growing.

Sea spiders can range in size. Some are less than 1 inch (2.5 cm) long. The largest are about 28 inches (71 cm) long.

That's WEIRD!

Some small sea spiders can walk on water!

SO MANY MORE

Other animals can lose different parts of their bodies to survive. The cotton rat can let go of the skin on its tail. That's helpful if they're picked up by the tail. The spiny mouse can let go of the skin on its back for the same reason. Maybe the weirdest of all is a kind of sea slug that can detach its head—and grow a new body from it!

These and other adaptations help many kinds of animals against predators and harsh habitats in so many ways. What other adaptations help animals survive?

This is the *Elysia marginata* sea slug. It can regenerate a whole new body from its head!

GLOSSARY

arthropod: An animal that lacks a backbone and has a skeleton on the outside of its body, such as an insect, spider, shrimp, or crab.

camouflage: Colors or shapes on animals that allow them to blend with their surroundings.

distract: To take attention away from something.

energy: Power used to do work or to live.

gene: Tiny part of a cell that is passed along from one living thing to its offspring.

gill: The body part that ocean animals such as fish use to breathe in water.

habitat: The natural place where an animal or plant lives.

hemisphere: One-half of Earth.

identical: Being the same.

marine: Having to do with the sea.

molt: To leave behind an outer covering that has become too small.

spinal cord: A thick rope of nerve tissue extending from the brain.

suction cup: A round cup that is fixed to a surface when pressed against it.

FOR MORE INFORMATION

Books

De la Bédoyère, Camilla. *Curious Features of Extraordinary Creatures*. New York, NY: Union Square Kids, 2024.

Garland, Michael. *We're Not Weird: Structure and Function in the Animal Kingdom*. New York, NY: Holiday House, 2023.

Websites

Axolotl

www.nationalgeographic.com/animals/amphibians/facts/axolotl

National Geographic provides facts about these water-dwelling creatures and what's being done to save them.

Sea Stars

aqua.org/explore/animals/sea-stars

Read what the National Aquarium has to say about sea stars.

Publisher's note to educators and parents: Our editors have carefully reviewed these websites to ensure that they are suitable for students. Many websites change frequently, however, and we cannot guarantee that a site's future contents will continue to meet our high standards of quality and educational value. Be advised that students should be closely supervised whenever they access the internet.

INDEX